Life Among the Milkweed

Buffalo Arts Publishing

Life Among the Milkweed. Copyright © 2021 by Melissa Borowicz Betrus. Printed in the United States of America. All rights reserved. No part of this book may be reproduced or transmitted in any form or by any means without written permission of the author. For information, address Buffalo Arts Publishing, 179 Greenfield Drive, Tonawanda, NY 14150.

Email: info@buffaloartspublishing.com

Cover and inside photographs by Melissa Borowicz Betrus

ISBN 978-1-950006-18-2

For Christopher,
Benjamin, and Parker

Life Among the Milkweed

By Melissa Borowicz Betrus

Author's Note

Life Among the Milkweed is a photographic showcase of the life that hides among the plants and grasses in the fields around our feet. This was a project begun out of my love for the monarch butterfly. Aware of its dependence on the milkweed plant, I stalked the insect there, and in doing so, discovered an unassuming mini-world of color and intrigue existing around me in silent acceptance of its place among human giants.

Common Milkweed
(*Asclepias syriaca*)

𝓘n a field of
milkweed,
secrets hide.
Pause.
Look.

Meet the gaze of a red milkweed beetle
whose black beady eyes peer over a leaf's edge
looking for danger before, in false death,
it drops down to safety among the grasses.

Red Milkweed Beetle
(*Tetraopes tetrophthalmus*)

Stare down the stink bug who stands its ground,
a shield-shaped warrior in a coat of armor
that if cracked by any foe,
would send the unpleasant smell of defeat
floating among the breezes.

Left: Green Stink Bug nymph
Right: Green Stink Bug adult
(*Chinavia hilaris*)
Far Right: Brown Stink Bug adult
(*Euschistus servus*)

Follow the paths of the ladybugs,
painted in various high-gloss shades,
detailed with patterned spots or random order dots,
as they scurry from stalk to leaf
rushing to satisfy an appetite for aphids.

Multi-colored Asian Ladybug eating aphid

Multi-colored Asian Ladybug Larva
(*Harmonia axyridis*)

Ladybug Pupa

Multi-colored Asian Ladybug Adult
(*Harmonia axyridis*)

Seven-spotted Ladybug
(*Coccinella septempunctata*)

Three-banded Lady Beetle
(*Coccinella trifasciata*)

Checker Spot Ladybug
(*Propylea quatuordecimpunctata*)

$\mathcal{S}$tudy the chaos on a seed pod
where milkweed bugs, infant to adult,
gather for a family picnic,
a group of red-orange gems
pushing head against thorax against abdomen.

$\mathcal{L}$ook again.

Large Milkweed Bugs
(*Oncopeltus fasciatus*)

*N*otice the single, oblong egg
    of the monarch butterfly,
      a tiny, cream colored droplet
left on the underside of a vast green canvas.

Watch in slowly moving moments
as a miniscule mouth makes a hole,
an escape hatch into the world
where it pulls its body
then turns to eat its one-time home.

*L*isten to the feasting of the caterpillar
carving crescent moons into the milkweed leaves,
head nodding in measured movements repeated
until the body is filled to bursting
and the tight, striped skin shrugged off
in growth spurts leaving it doubled in size.

Wonder at the mystery happening
inside the shining emerald chrysalis
until time fades its color to clear
revealing a black, blood-fattened body,
a wisp of silky orange wing.

Observe the birth of the butterfly
whose legs anchor it to safety
as its bloated body pops from its protective casing

and it dangles, pulsing blood into its wings
until they reach their full splendor
and are ready to beat the air.

Monarch Butterfly emerging
(*Danaus plexippus*)

Linger in the magic of the monarch
tiptoeing across the firework blooms of the milkweed
as it unfurls its proboscis to sip the sugar-sweet nectar,
offering kisses of thanks before lifting from its host
and leaving it behind.

Watch it go,
then return to the milkweed.

Left: Milkweed Leaf Beetle nymph
Right: Milkweed Leaf Beetle eggs
Middle: Milkweed Leaf Beetle eggs hatching
Far right: Milkweed Leaf Beetle adult (*Labidomera clivicollis*)

$\mathscr{S}$it with a freshly pinked and hunch-backed larva
with dashed-line decorations bolded
on a slick accordion skin
not yet tough enough to promise protection.

Delight in the colors of the candy-striped leafhopper
boldly stroked by the brush of a master make-up artist,
a model who walks a milkweed runway
then launches to another stage
to await the oohs and ahhs of another audience.

Right: Candy-Striped Leafhopper
(*Graphocephala coccinea*)

Upper left: Rhododendron Leafhopper
(*Graphocephala fennahi*)

Admire
the strength of the grasshopper,
whose crimson-colored legs,
held together with parts like brass fasteners,
hug tight the plant it hopes will conceal its sectioned shell
while with jaws set to serious, it stares
from its bulging, oval eyeballs.

Two-striped Grasshopper
(*Melanoplus bivittatus*)

Marvel at the bristled-bodied caterpillar,
a miniature scrub brush in a costume
of skunk and tiger stripes
content to bend its body to the contours
of a leaf slightly browned and slowly dying.

Above: First, Second, and Third Instars of the
Milkweed Tussock Moth Caterpillar
Right: Adult Milkweed Tussock Moth Caterpillar
(*Euchaetes egle*)

Glimpse

    the golden-ringed eye of the spring peeper,
sitting in silence against a stalk
until the sun spreads pastels across the sky,
when it will stretch the skin at its throat
and soothe the night with a whistle-song serenade.

Spring Peeper
(*Pseudacris crucifer*)

In a field of milkweed,
secrets hide.

Milkweed and the Monarch Butterfly

Milkweed is the host plant to the monarch butterfly. It is the only plant on which it will breed, and it is the plant on which the monarch caterpillar is totally dependent.

While many insects have a variety of food sources, milkweed is the only food source for the monarch caterpillar. Without it, the caterpillar is unable to develop into a butterfly.

Research conducted through organizations like Monarch Watch and the World Wildlife Fund shows that monarch populations are declining due to the destruction of milkweed areas. In January 2013, the World Wildlife Fund declared the phenomenal monarch migration of the eastern monarch butterfly population to Mexico as endangered. However, this does not have to mean the end of the monarch butterfly.

Saving the monarch involves restoring monarch habitat, and the answer to how is simple: plant milkweed. Create a dedicated space and carefully tend your plot. Scatter the seeds in a game of chance and let nature be their keeper. Float the seeds on an autumn breeze with a wish of luck and hope. Just plant it. Plant it in any space, in any soil. Plant it, and your reward will come on the strength and dreams of orange and black wings.

For more information, please visit Monarch Watch (www.MonarchWatch.org), Monarch Joint Venture (monarchjointventure.org) or Journey North (journeynorth.org). I am in no way affiliated with these organizations. I am simply a heart that has been captured by the brilliance of a butterfly.

Made in the USA
Monee, IL
30 October 2021